IT Troubleshooting a Beginners Guide

Written by Joe Blanc

This is the 2nd edition of this book and

is dedicated to all my friends and family who supported me.

This book is meant to be a guide for the beginners in IT who just graduated or are looking for their first technology job. It is meant as a guide to better understand the importance of good troubleshooting to fix the most common IT problems.

I have been working on contract here in Canada for 20 years and I have done everything from server support, helpdesk, desktop support, onsite technician as well as general tech support.

The one thing that I found, is that troubleshooting was never offered as a supplement course to the MCSE but I did learn basic troubleshooting in my Cisco CCNP course that I took. As well as earning those two certifications I also became an A+ computer technician.

I earned all these certificates in the early 2000's and they led me to almost constant work since then.

So what this book is about, is based on those years of working and the courses that I completed.

Getting the first Call

After going through the interview process and testing and landing your first job in IT, most people will find themselves in a call center or helpdesk.

With a little training on proprietary systems within the company you are expected to answer calls pertaining to technical problems with systems, software, etc...

It is of utmost importance that you start taking your calls with confidence and an air of customer service. Upon answering your first call you must keep in mind your training and not to stray from that base of knowledge. There are a few points to follow:

1) Greet the person calling and make sure you listen carefully.

2) Address the problem whether software or hardware if it is unique to the caller or widespread.

3) If local to caller troubleshoot in normal way.

4) If problem is widespread make a support ticket to the appropriate department.

These are the general points that you will come across time and time again. There are specific rules to every company on timing, such as SLA(service level agreements) made, so you have to follow procedures carefully.

As far as troubleshooting, on the phone for local problems, there are a few key points to consider:

1) Is this the first time this has happened.

2) Has there been any updates that have gone through and check version of software.

3) Does software need internet connection to work.

If a hardware problem:

1) Is it an intermittent problem or not.

2) Is your work affected by this problem or not.

Depending on the problem a lot of the questions will explain themselves once you hear the description of the problem.

It is always important to let the caller know the time frame when they can expect a fix or how you will proceed to fix their problem.

Once the call is over and you've put the support ticket in the system you can get ready for your next call.

Busy call centers without any outage can get 50-70 calls a shift. Of course there are slow days as well as call centers that experience no more than 10-20 calls a day

Desktop Support the Next step

Some helpdesk jobs will require you to do desktop support. There are as well, strictly desktop support position that you will no doubt apply for. In these jobs, you are required to know hardware as well as software support. You will be called upon to support workstations, printers, projectors, etc...

Your day will be filled with support tickets passed on to you by the helpdesk, to fix more time consuming problems or problems that require you to visit the caller in person.

At most places you will have no more than 20 tickets a day to take care of and most but not all problems will be time consuming. You will also most likely be responsible for new user installs and imaging of new laptops or desktops.

Always approach every day here with a sense of urgency because normally by the time you get the ticket and respond to the caller it could be an hour or two since the caller originally called and most will be anxious to get their problem solved.

Also, every company has VIP callers that have no time or inclination to troubleshoot with the helpdesk and require someone almost immediately to respond to there problem. In cases like this it is usually a helpdesk person or your manager that will alert you to the call and it is advisable to drop everything and respond to the caller most of the time in person.

Now that you have an idea of what to expect and what will be required of you in your first jobs it's important to look at what the steps of troubleshooting are.

I mentioned some basic helpdesk steps but when it comes to desktop support you will be faced with more complex problems. You will have to use all of your training and sometimes that's not enough. In cases such as this, I recommend researching the problem on the internet at manufacturers websites as well as software support websites, if it is a problem you have no idea about how to solve. If after researching the problem you still can't find a solution then I would ask one of your senior colleagues at the office if they have seen the problem before or know of a way to fix the problem. They will be more willing to help if you explain you have researched and have not found a solution.

Troubleshooting can be defined as finding the root cause of the problem and then implementing a solution. Whether it's hardware or software there is always a root cause and once you understand the problem you can implement a fix.

1) Computer hardware problems usually require a hardware fix. Whether intermittent or constant you will have to find parts from your supplier.

2) Software problems usually are fixed with updates, settings or re-installs in the case of local issues.

3) Printers problems usually require printer parts or could be related to cabling or software such as faulty driver.

4) Network problems locally are usually related to the network card or cabling and sometimes the jack in the wall has been turned off.

What is important is always to find the root cause and then apply the solution.

Server Support

Server support both on the phone or installations or repairs are usually handled by personnel with several years experience. Because of the mission critical applications and services the server provides it's essential that they be fixed right.

For new installations of server systems, you can expect to work onsite for awhile while you unpack the hardware and assemble anything that needs to be assembled or connected. Then in the case of servers all operating systems need installation and configuration as well as any system software or application servers have to be installed and configured.

In some cases, while you are onsite you may have to connect and configure routers and switches depending on the service that is being installed.

If you have a position as application server support, you will be faced with monitoring local application client problems on workstations and how they relate to the server. Sometimes these problems are solved by the helpdesk or desktop support but alot of the time, especially with new systems the problems are routed to server support.

The key is to isolate the problem to whether it is the client or server that is the root problem and go from there. You can usually check the support ticket to find out what has been done already to resolve the problem. Sometimes it involves a hardware issue or problem with the application software that needs an update or change of settings.

As far as phone support, you will be entrusted with quick fixes or sending a technician onsite to further fix the issue. SLA's will be key as most server problems will be critical. In some cases you will have to connect remotely to the server and check settings or software if it is not working properly. Also, with the help of the caller you can identify hardware issues such as a RAID drive failure or if proper lights are on a router to indicate a failure or not.

Repair and installation technicians will go onsite and replace hardware or install systems from scratch.

I've had to go onsite to swap hard drives, monitors, router's as well as install applications etc...

General Technical Support

Tech support can be a rewarding job to say the least. You usually work in a big call center but not always. You spend your time answering calls from customers and spend your eight hours or sometimes more fixing problems remotely. By using software onsite or remote desktop software you will help customers and solve there problems whether related to the company's service or the customers computer system.

In most cases you will be trained on the company's proprietory applications and services and you will then answer calls based on that training. Calls will range from irate callers to people just wanting more information on the service the company you work for provides.

All call centers require you to be on time for your shift and to be very responsible for the company's image, as you represent the company on every call.

Usually but not always it involves an eight hour shift. Alot of technology companies as well as telecommunications companies offer very lucrative positions which involve basic fixes as well as sending a technician to there clients businesses.

The kind of work that you will be doing will be email support, service support and in the case of telecommunications companies being aware of service outages.

These types of positions are very similar to helpdesk positions in that you will be getting around 50-70 calls a day on busy days. If you work the overnight shifts usually the call volume is greatly reduced and on days that there are no major problems you can expect to get 20-30 calls a shift.

With all of these technical jobs comes responsibility's and discipline to get the job done. Hopefully this book gives you a better understanding of troubleshooting computer systems and what you will be required to do while you work in IT.

The End

www.ingramcontent.com/pod-product-compliance
Lightning Source LLC
Chambersburg PA
CBHW071203220526
45468CB00003B/1147